Introduction

The aim of this book is to provide a course in the interpretation of West Indian topographical maps for students who are preparing for O-level and similar examinations. It may therefore be used as a companion volume to *Caribbean Lands* which deals in a descriptive way with the general and regional geography of the area at about the same level.

Though *Map Reading for the Caribbean* is not intended for absolute beginners, it does take into account the fact that many students who are familiar with the elements of geography have had little opportunity to relate these elements to the West Indian landscapes represented by the range of maps contained in this book. This range is considerable—a young volcanic landscape (Montserrat); older volcanic areas (Southern Antigua and Eastern Grenada); folded mountain foothills (Southeastern Jamaica); a terraced limestone area (Barbados); a flat clay coastal plain (Guyana); and an interior alluvial plain (Trinidad). In addition, a variety of coastal and valley forms are represented.

Map Reading for the Caribbean is a workbook based on the principle that the best way to learn how to interpret maps is to carry out a series of exercises based on them. Of the 47 assignments contained in this book, 35 follow a carefully graded sequence designed first to teach the basic skills of map reading and then to assist students to use what they have learned in interpreting features of increasing complexity. Thus the book progresses from the simple skills of measuring directions and distances, to the teaching of various methods of describing and interpreting physical landscapes (including the calculation of gradients and the drawing of sections), the description of physical regions, the significance of certain geographical relationships, and the human and social geography of selected areas. Model examples are frequently used to introduce new concepts, and by the time a student has completed the course he should be sufficiently familiar with the methods of map interpretation to be able to tackle any West Indian map on his own, without further assistance. He will also have gained an insight into some of the ways in which the geographical skills he has mastered can be of practical value in the world in which he lives.

The remaining twelve assignments have been included as revision exercises to ensure that students do not forget the techniques they have learned or allow their speed of map reading to lapse. As for the three tests at the end, these have been added to help students to prepare for examination conditions.

Estimating and measuring directions

In order to help map readers to judge directions quickly and easily, most maps are printed so that North is at the top. This is true of all the maps in this book, though to read some of them (Maps 3, 6 and 7) you will need to turn the book sideways. Look for example at Map 1 (page 15). On this map, estimating directions is particularly easy because squares (known as a grid) have been ruled over it. All the lines ruled from the top to the bottom of the map point in a North to South direction. All those ruled across the map point from West to East.

Figure 1 shows a similar grid. Write the correct directions in the empty circles.   Assignment 1

Figure 1

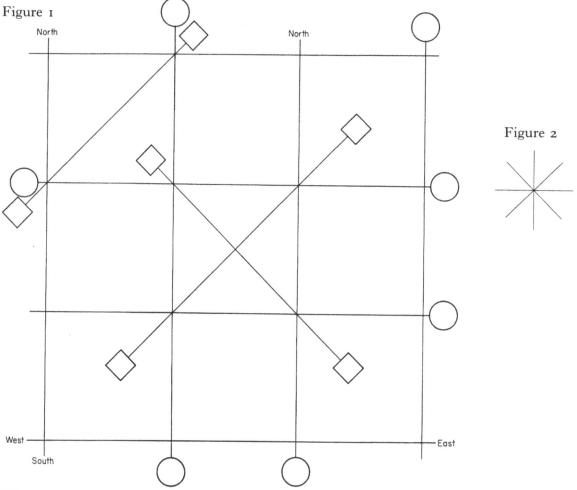

Figure 2

Diagonals across the squares give NE, SE, SW, and NW.

Write the correct directions in the empty diamonds in Figure 1.   Assignment 2

Write the correct directions in Figure 2.
These eight directions are occasionally sufficient to describe directions accurately. For example on Map 1 you can see that Gages Upper Soufriere is South of the tower at Rileys Estate. At other times these directions are close enough to being correct that they may be said to be *approximate*.   Assignment 3

(a) Look at Map 1 (page 15). What direction is the tower at Rileys Estate from Gages Assignment 4

Upper Soufriere? ....

(b) What direction is Gages Upper Soufriere from Castle Peak?...........................

and from △M15? (in square 11J) ....

(c) What direction is △M15 from Gages Upper Soufriere?...............................

Many directions cannot be stated accurately in this simple way, though sometimes they can still be described in words. For instance, the direction midway between North and North-East is North-North-East, and the direction midway between North-East and East is East-North-East. These directions are shown in Figure 3.

Figure 3

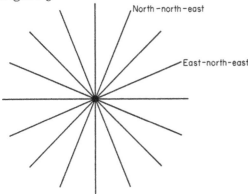

Figure 4

Name the other directions in Figure 3 Assignment 5

If, as is often the case, the direction of one place from another cannot be stated sufficiently accurately by one of the sixteen names given in Figure 3, it is better to give a compass bearing than to subdivide them again in terms of such complicated words as North-North-East-by-North. Compass bearings are measured in degrees read *clockwise* from North. Thus, because East is 90° in a clockwise direction from North its bearing is 90°. Some of the other bearings of the eight major directions are shown in Figure 4.

(a) Complete Figure 4 giving directions and bearings. Assignment 6

(b) Here is a bearing of 346°

Is it closest to

(i) North-North-West (ii) West-North-West (iii) North? ....

Using Map 1, the example 'What direction is the tower at Whites Estate (square 10K) from the church at Harris Village (square 9J)?' may be used to illustrate the method of finding the compass bearing of one place from another.

On Map 1 rule with a sharp pencil a north-south line through the place you are Step 1
starting from (in this case the church at Harris Village).

Rule another straight line between this place and the place for which you want to find Step 2
a bearing (in this case the tower at Whites Estate.)

At the place you are starting from, measure with a protractor the angle between North and the line joining the two places, as in this illustration.

Figure 5

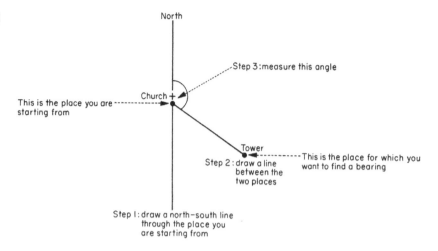

(a) The bearing of the tower at Whites Estate (Map 1) from the church at Harris Village is .... The bearing of the church at Harris Village from the tower at Whites Estate is...
(b) Explain why the difference between these bearings is 180°.

....

....

(c) What is the bearing of the church in square 10J from the tower at Farrells Estate (square 8K)? .... from the centre of Gages Upper Soufriere? ....

and from the ruined tower at Mulcares Estate (square 11L)? ....
(Remember always to measure clockwise from North.)

(d) From M14 the bearing of M8 is 150°, that of M4 is 189°, and that of M9 is 234°. Without looking at the map, state the bearing of M14 from M8 ...., from M4 .... and from M9 .... Now use the map to check your answers.

(e) Look at the map and estimate the approximate bearings of the tower at Farrells Estate from the ruined tower at Mulcares Estate .... and from the church at Harris Village .... Now check your answers.

NOTE: Now that you know directions, be careful how you speak about the top and bottom of a map. Otherwise you might one day fall into the trap of saying 'Chicago lies at the bottom of Lake Michigan'.

Scales and distances

The topographic maps in this book have all been drawn in proportion to the parts of the West Indies they represent. In other words the distance between any two points on a map is a definite proportion of the true distance between those places on the ground. This proportion is the scale of the map.

The scale of Map 1 is 1:25,000. This means that one inch on the map represents 25,000 inches on the ground. So if it were possible to enlarge the map 25,000 times it would exactly cover the part of Montserrat which it portrays.

It is worth remembering (or being able to work out quickly) that 1 mile equals 63,360 inches. Then you can work out that if a map has a scale of 1:25,000, one inch on the map represents $\frac{25,000}{63,360}$ of a mile. Put as a decimal, this equals 0·395 of a mile. Or, as a fraction, an inch on the map is about $\frac{4}{10}$ of a mile on the ground.

(a) Complete this table for a map with a scale of 1:25,000

Assignment 8

TABLE 1

ON THE MAP Number of inches	ON THE GROUND Approximate number of miles
1	$\frac{4}{10}$
2	
3	$1\frac{2}{10}$
4	
5	
6	

(b) About how many inches on Map 1 represent 2 miles on the ground?

....

(c) About how many inches on Map 1 represent 20 miles on the ground?

....

(d) Working to three decimal places, how many inches on Map 1 represent 20 miles on the ground? ....

(e) What difference is there between your approximate answer (c) and your accurate answer (d)? ....

(f) For what purposes do you think your approximation would suffice?

....

....

(g) For what purposes do you think it would not be accurate enough?

....

....

....

Look at Map 3 on page 23. Remember that North is at the top of the map (not at the top of the page). Thus grid line 37 runs in a West-East direction.

(a) Underline the correct word (or words) in brackets.
 (i) Bachelors Hall is (North/South/East/West) of Bowden
(ii) Places south of Hordley are (Bath/Hampton Court/Holland Bay/Rocky Point/ Bachelors Hall)
(b) The bearing of Pera Point from the Police Station at Port Morant is

.... and that of the Police Station from Pera point is therefore ....
(c) On this map how many inches represent 20 miles on the ground?

....

Though it is possible to find the distance between places by calculation as in ASSIGNMENT 8, there are easier ways of making the measurement. Here is the way to find the shortest distance (i.e. the distance in a straight line) between two points.

Lay the straight edge of a piece of paper on the map so that it touches both points. Mark the two points on the paper using a sharp pencil.

Transfer the marked paper to the graduated scale line accompanying the map and read off the distance in miles and parts of a mile.

Turn to page 14 and Look at Linear Scale A which accompanies Map 1. It is marked so that distances can be read in fractions of a mile and in hundreds of yards. Note how the parts of a mile are shown to the *left* of the zero position.
(a) What is the direct distance between △ M16 (square 7J on Map 1) and △ M7

(square 8H) (i) to the nearest mile? ...., (ii) the nearest half-mile? ....,

(iii) the nearest hundred yards? ....
(b) Estimate and then measure as accurately as possible the direct distances between the places shown in this table.

TABLE 2

PLACES	ESTIMATION (without measuring)	MEASURED ACCURATELY
△ M16 to △ M17 (square 8J)		
△ M16 to △ M14 (square 10L)		
△ M16 to the nearest point on the coast		
△ M16 to the nearest main road		

In cases where straight line distances are not sufficient, it is possible to mark the point and twist the paper when you come to a corner. Better still, you can lay a length of thread along an angular or curved line, and measure it against the scale.

An alternative method is to use dividers. Here is an example of the way they may be used to measure the length of a curving road (i.e. the distance AE in Figure 6).

Figure 6

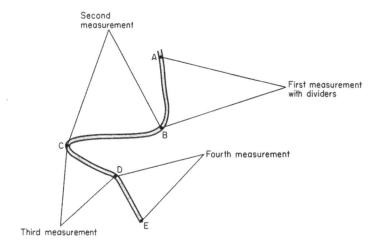

Rest one of the points of the dividers lightly on the map at A and the other point at the place where the road makes its first major bend (i.e. at B).

Step 1

Take a piece of paper and prick it with both points of the dividers. You have now transferred the length AB from the map to the paper. Rule a line between these points and extend it for what you estimate to be at least the distance AE.

Step 2

Repeat Step 1 for the next fairly straight stretch of road (BC). Transfer this length to the ruled line on the paper by resting one point of the dividers in the hole already marked at B and pricking a new hole (C).

Step 3

Repeat the process for lengths CD and DE. The row of holes in the piece of paper will now look like this:

Step 4

Figure 7

Open the points of the dividers until they touch holes A and E. Transfer them back to the map and rest them lightly along the graduated scale line. Read off the distance in miles and parts of a mile.

Step 5

(a) How long is the coastline shown on Map 1? ....

Assignment 11

(b) How far is it from the end of the road at Paradise Estate (square 9K) to the road

junction at Long Ground Village (square 10L) in a straight line? ....

and by road? ....

(c) How many miles of motorable road are shown on the map? ....
(d) Dividers are very useful to prevent the eye from being deceived when measuring the direct distance between two points. Use them to check your measurements in

TABLE 2 on page 6. If your answers do not agree, explain your error. ....

....

Three ways of describing and depicting landscapes

It takes practice to learn to visualize a landscape depicted by a topographical map. To help you to gain this practice, three types of exercise are given in this book.

TYPE 1

Imagine you are standing on a beach. Describe the sort of landscape you would cross if you were to walk a short distance inland. For example if you walked ½ mile inland from the east coast of Montserrat along the line shown on Map 1 which separates squares 11J and 11K your description would be something like this: 'A steep climb up to a height of 200 feet is followed by a gentler ascent to between 250 feet and 300 feet. There is then a descent of about 50 feet to the bottom of a narrow valley in which a small stream flows in a north easterly direction. After that the land rises again, and half a mile from the sea it is between 400 and 450 feet above sea level.'

Assignment 12

(a) Describe the landscape you would cross if you were to walk ¾ mile inland from the east coast of Montserrat (i) along the line separating squares 11I and 11J.

....

....

....

....

(ii) along the line separating squares 11N and 11O (that is, in the extreme south of the area shown on Map 1).

....

....

....

....

(b) Describe the landscape you would cross if you were to walk ¾ mile due north from the coast at Falmouth (Map 2 on page 19) along a line which would take you beside the west wall of St. Pauls Church.

....

....

....

....

TYPE 2

Colour the highland and the lowland so that the map resembles a relief map in an Atlas. This makes the map much easier to understand, but it also gives the impression that the land rises by a series of steps which, you should remember, is not really the case.

Assignment 13

Colour the land shown on Map 2 which is above 300 feet *light brown*. Colour the land below 100 feet *light green*. Colour the land in between *yellow*.

Draw a section. This gives the impression that the land has been cut vertically in two with a saw and makes it easy to visualize. The procedure is as follows:

TYPE 3

Lay the straight edge of a strip of paper on the map along the line selected for the section. With a sharp pencil mark and number the points where the contours disappear below the paper. (You will find this becomes easy with practice and that you do not have to number all the contours to get a clear picture of the landscape.)
Figure 10 (on page 10) shows this step of the procedure for a section drawn along the lines separating squares H and I on Map 1.

Step 1

Lay the strip of paper beneath a series of parallel lines like those shown in Figure 8. (Usually these lines are drawn $\frac{1}{10}$ inch apart and represent an interval of 100 feet.) Make a faint pencil dot on the appropriate line above each mark on the paper strip.

Step 2

Figure 8

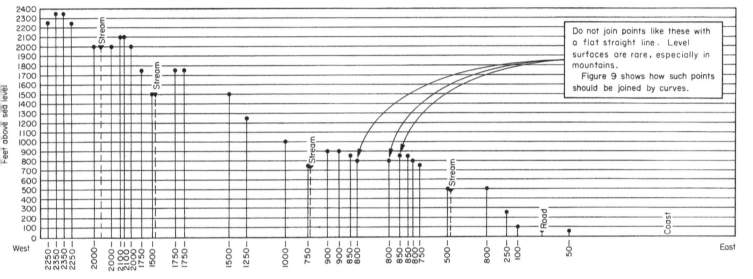

Do not join points like these with a flat straight line. Level surfaces are rare, especially in mountains.
Figure 9 shows how such points should be joined by curves.

Join the dots with a pencil in a *curved* line, using your judgement to decide the heights of hill tops and valley bottoms if their exact heights are not known. If you wish you may shade in everything below this line to represent land. Name each significant feature of the landscape, and give the section a scale and a caption. This result is shown in Figure 9.

Step 3

Figure 9   *Caption:* SECTION DRAWN FROM WEST TO EAST ACROSS PART OF MONTSERRAT

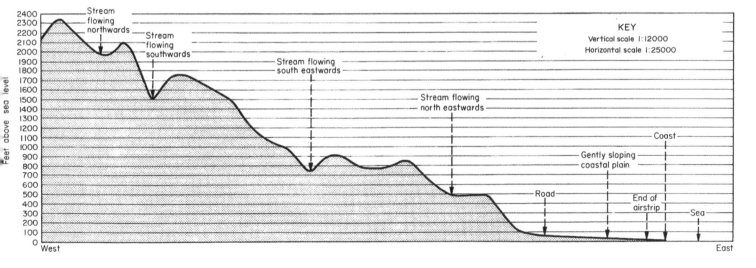

KEY
Vertical scale 1:12000
Horizontal scale 1:25000

It is important to include a scale because the horizontal and vertical scales of a section are seldom the same and it is necessary to know to what extent the drawing represents the true landscape. In the example given in Figs. 8 and 9, the vertical scale chosen for the section is $\frac{1}{10}$ inch representing 100 feet. Therefore 1 inch represents 1,000 ft., or 12,000 inches, and the scale is 1:12,000. The horizontal scale is of course the same as that on the map, that is 1:25,000. So the slopes shown in the section are approximately twice as steep as they are in reality on the ground. There is nothing wrong in exaggerating the vertical scale of a section to this extent. In fact in depicting areas of gentle relief it is necessary to do so if changes of slope—which so often provide clues to the identification of particular landscape features—are to be shown clearly. However, in all such cases it is essential to state the extent of vertical exaggeration, and one of the best ways of doing so is to write both the vertical and the horizontal scales below or beside the section.

(a) Using Map 1 draw a section from Gages Upper Soufriere eastwards to the sea.   Assignment 14

KEY

Figure 10

(b) Using Map 2 draw a section from St. Pauls Church, Falmouth, to the end of the 'mineral railway' at Bodkins. (In fact, the railway was built to carry sugar cane). Note that on this map the Vertical Interval (V.I.) between the contours is not the same throughout. Below 100 feet the V.I. is 25 feet: above 100 feet the V.I. is 100 feet. Changes in the Vertical Interval occur in most Western Indian maps, so be sure to check the contours carefully before attempting to draw a section.

Caption : ....

```
700
600
500
400
300
200
100
  0
```

Key :...

....

(c) Using Map 2 draw a section southwards from the point where the parish boundary leaves the northern tip of Christian Cove to the head of Indian Creek.

Caption : ....

```
300
275
250
225
200
175
150
125
100
 75
 50
 25
  0
```

Key :...

....

(d) What approximately is the vertical exaggeration of this section? ....

(e) Why has a Vertical Interval of 25 feet been chosen to portray the landscape of Antigua below 100 feet? ....

....

....

....

(f) Why has a Vertical Interval of 100 feet been chosen for the higher land?

....

....

....

....

(g) Estimate the heights of (i) Fort Charlotte ...., (ii) Fort Cuyler ....,

(iii) Old Government House ...., (iv) Clarence House ....

What is the most you could be wrong? ....

(a) How long is the airstrip at Blackburne Airfield (Map 1)? (i) In yards ....

(ii) In metres .... In what direction does it lie? ....

(b) What do you estimate the shortest distance between St. Pauls Church, Falmouth

and Ffryes (Map 2) to be? (i) In yards .... (ii) In kilometres ....

Measure the true distance. What is it? .... yards, ... kilometres.

(c) Estimate the distance in km. in a straight line from Chapel Point to Charlotte

Point (Map 2) .... Measure it. What is the true distance? ....

(d) Estimate the length of the coast in km. from Chapel Point to Charlotte Point

.... Measure it. What is the true distance? ....

(e) What is the compass bearing of Ffryes from St. Pauls Church? .... How

many degrees is this away from North-East? ....

(f) Without measuring, state the bearing of St. Pauls Church from Ffryes ....

(g) Colour lightly all the land shown on Map 1 which is over 2,000 feet purple, that between 1,500 feet and 2,000 feet red, that between 1,000 feet and 1,500 feet brown, that between 500 feet and 1,000 feet yellow, and that below 500 feet green. Complete the colour key on page 14.

(h) Describe the landscape you would cross if you were to walk inland from Conference Point (Map 4 on page 27) along the road to the school at Tivoli.

....

....

....

....

....

....

....

....

....

....

....

....

....

....

....

....

....

....

Giving a grid reference

Jamaica is one of the few West Indian territories whose maps are covered with a numbered grid. This makes it possible to refer to any point on one of its maps simply by means of a six-figure number. The procedure is best explained by using an example.

'State the grid reference of the church at Golden Grove (Map 3).'

Look along the bottom of the map on page 23 and find the number at the western edge of the square in which Golden Grove Church lies. This number is 80. — Step 1

Imagine the square from 80 to 81 to be divided into ten equal parts. Estimate the number of these parts that occur between line 80 and the church. The answer is about 3 parts. (If the church were in the middle of the square the answer would obviously be 5 parts.) — Step 2

The first three figures of the grid reference are therefore 803.

Now look along the side of the map. Find the number at the southern edge of the square in which the church lies. This number is 37. — Step 3

Estimate the number of tenths northwards. This figure is about 6. The last three figures of the grid reference are therefore 376. Putting the two together the full six-figure grid reference is 803376. — Step 4

NOTE: Always keep to this order of procedure giving the 'eastings' (the easterly part of the reading) before the 'northings' (the readings northwards).

Look at Map 3 and give six-figure grid references for (a) Holland Great House .... — Assignment 16

(b) the Post Office at Dalvey .... (c) Pear Tree River School ....

(d) Bath Police Station .... (e) Pera Point ....

(a) Which of these answers is correct? On Map 3, with a scale of 1:50,000. 1 inch — Assignment 17 (Revision)

represents (i) 1·267 miles (ii) 1 mile (iii) 0·789 miles ....

(b) Estimate the straight line distance from Bath Botanical Gardens to the mouth

of the Plantain Garden River .... What is the true distance? ....

(c) How long is the river itself between these two points? ....

(d) If a ship is 2 miles south of the Post Office at Dalvey, what is its grid reference?

.... From the ship, what are the bearings of

(i) Snook Point? .... (ii) Rocky Point? ....

(e) Look carefully at Map 3. What heights are the contours? ....

....

What, therefore, are the Vertical Intervals on this map

(i) between sea level and 100 feet? ....

(ii) between 100 feet and 250 feet? .... (iii) over 250 feet? ....

MAP 1 Part of Montserrat

Main motorable Roads .... Bridge

Secondary motorable Roads ....

Other Roads and Tracks (non motorable) ....

Footpaths ....

Named Buildings ....

Police Station .... ■ Pol Sta

Post, Telegraph Office .... PT

Hospital .... ■ Hosp

Church, Chapel, School .... ■ Ch ■ Cha ■ Sch

Court House, Hotel .... ■ Ct H ■ H

Fences, Walls ....

Water Tank .... ■ WT

Trigonometrical Stations:- *Main* .... △

Trigonometrical Stations:- *Minor* .... ▽

Heights in feet above sea level 101

Contours (Vertical Interval 50') 300 250 200

Forest, Thicket and Scrub Woodland ....

Dispersed trees amongst Settlements ....

Hedges and Terraces ....

Cliffs, Screes and Land-slips ....

Plantations (Sugar Cane, Limes, Coconuts etc.) ....

Watercourses

SEE ASSIGNMENT (G)
ON PAGE 12

KEY

below 500ft	
500—1000ft	
1000—1500ft	
1500—2000ft	
over 2000ft	

Scale 1:25,00

Linear Scale

1 Mile   ¾   ½   ¼   0   1   2 Miles

Yards   1000   0   1000   2000   3000   Yards

Linear Scale

1 Kilometre   0   1   2   3 Kilometres

(f) Describe the landscape you would cross on Map 3 if you were to walk northwards from the coast to ▽ Bowden 293 (Grid reference: 790359).

....

....

....

....

....

....

....

....

(g) Draw a section from due south from the village of Spring Bank (782385) to the sea. Label it. Do not forget to add a caption and a key.

700
600
500
400
300
200
100
0

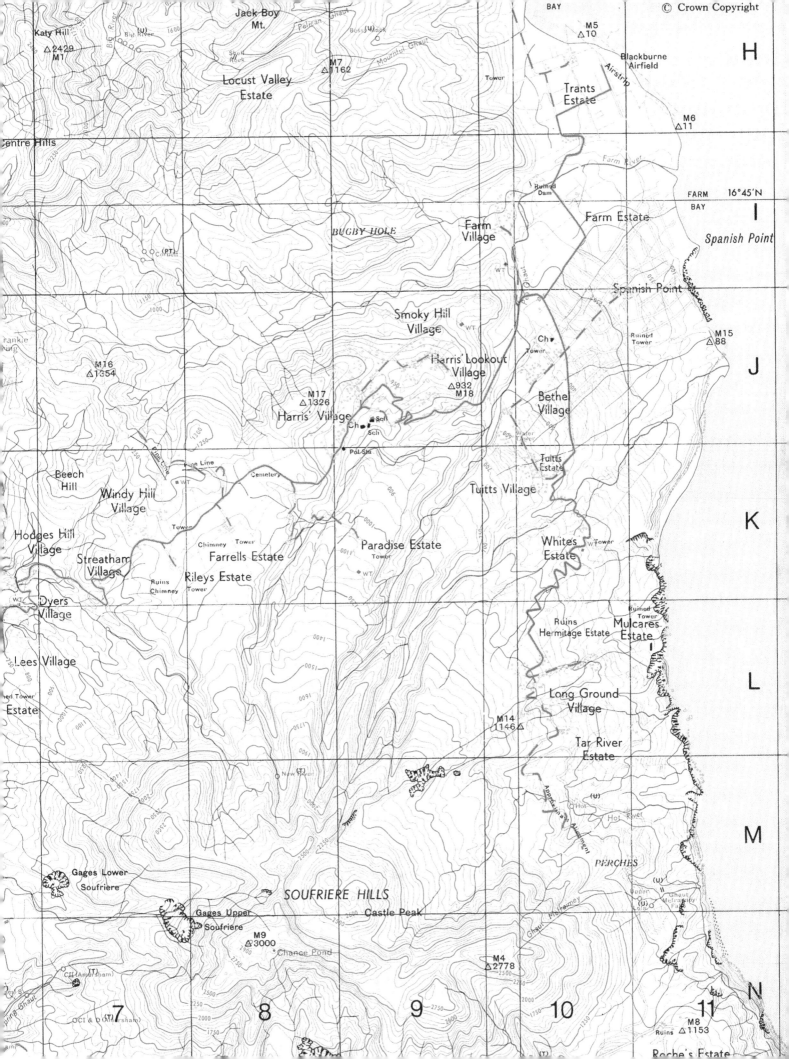

More about landscapes

With the knowledge of how to draw sections you can determine whether any two points on the ground are visible from each other. All you have to do is to rule a line joining the points where they appear on the section and see if any land intervenes. If it does not, the two points are visible from each other.

INTERVISIBILITY

Here, as an example, is the section originally shown in Figure 9. The line joining A to D passes above hill B. Therefore D (the coast) can be seen from A, and A can be seen from D.

Figure 11

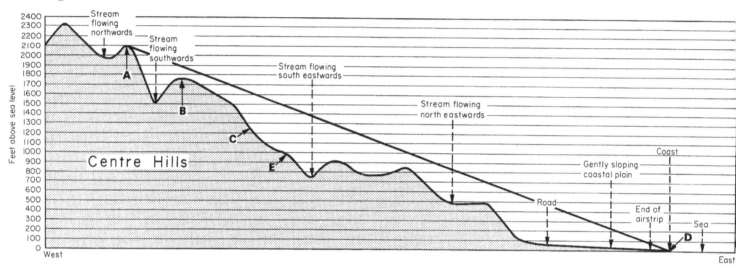

(a) Is point C visible from the coast D? .... (b) What is the lowest height of the Centre Hills which is visible from the coast? .... (c) About how high would you have to be in an aircraft above the coast to see point E? ....

Assignment 18
(using Figure 11)

A gradient is a way of expressing the steepness of a slope. It is worked out by dividing the difference in height between the top and bottom of the slope into the horizontal distance between them and expressing the answer as a ratio. In other words.

GRADIENTS
AND SLOPES

$$\text{Gradient} = \frac{\text{Horizontal distance (measured in feet)}}{\text{Difference in height (measured in feet)}}$$

Take as an example the line AD drawn in Figure 11. A is 2,100 feet high; D, being on the coast is 0 feet high. The difference in height is therefore 2,100 feet. The horizontal distance between them is about 4,000 yards, that is 12,000 feet. (Check this from the scale given on Map 1). The gradient is therefore the ratio:

$$1 : \frac{12,000}{2,100} = 1 : 5.7$$

The answer is expressed as follows: 'The gradient of AD is 1 in 6', gradients always being expressed to the nearest whole number.

A slope may also be measured in terms of its angle from the horizontal. The figures used to obtain this angle are the same as those for obtaining a gradient, but they are inverted and worked out as a decimal which is then converted to an angle by referring to a table of tangents. Thus, to use the same example as before,

$\frac{2,100}{12,000} = 0.175$ Tangent tables show that this corresponds to an angle of about 10°.

Here it is as a diagram A gradient of 1 in 6, that is a 10° slope.

(a) What is the average gradient of the slope AE shown in Figure 11? .... Assignment 19

(b) Given that the height of Plantain Garden River (Map 3) at Bath is 100 feet, what is the gradient of this river from Bath to its mouth? ....

(c) What is the gradient of the Cocoa River? (Map 3) ....
(d) State three differences between the Plantain Garden River and the Cocoa River.

(i) ....

(ii) ....

(iii) ....

(e) Using gradients as part of your answer, compare Monks Hill (Map 2), Castle Peak (Map 1), and the hill between St. John and Upper St. Johns (Map 4).

....

....

....

....

....

....

....

....

....

....

....

....

(f) Engineers must know the gradient of a river when they are making plans to build a dam across it. Give other examples in which a knowledge of gradient is necessary.

....

....

....

Soil conservation of some sort is necessary on all cultivated land with a slope of more than 5°. Slopes between 20° and 30° must be cultivated with extreme care. Those of 30° and more are very steep indeed, and if they are cultivated their soils are very likely to be washed away by rainstorms. Assignment 20

MAP 2 Part of Antigua

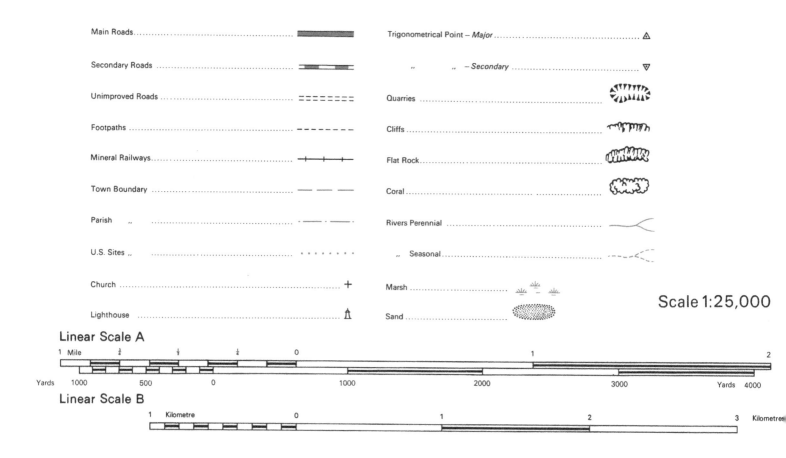

Main Roads	Trigonometrical Point – *Major* △
Secondary Roads	„ „ – *Secondary* ▽
Unimproved Roads	Quarries
Footpaths	Cliffs
Mineral Railways	Flat Rock
Town Boundary	Coral
Parish „	Rivers Perennial
U.S. Sites „	„ Seasonal
Church +	Marsh
Lighthouse	Sand

Scale 1:25,000

Linear Scale A

1 Mile ¾ ½ ¼ 0 1 2

Yards 1000 500 0 1000 2000 3000 Yards 4000

Linear Scale B

1 Kilometre 0 1 2 3 Kilometres

Using this as a guide, what degree of soil conservation would you recommend in the following areas?

(a) From Bethel Village (Map 1 square 10J) to the sea ....

(b) From Monks Hill (Map 2) south to Falmouth ....

(c) The land in the vicinity of Upper St. Johns (Map 4) ....

(d) The land in the vicinity of Upper Pearls (Map 4) ....

DESCRIBING COASTS

A coast is the strip of land which borders the sea. When you are setting out to describe one there are many things to consider. Here are some of them. The sea may contain islands or reefs sufficiently close to the shore to be worth mentioning. The shape of the coast should be stated; for example it may be smooth, gently curving, or deeply embayed. In character it may be cliffed or low lying; in composition it may be sandy, rocky, muddy, or marshy.

Here, as an example, is a description of the coast running southwards for about a mile from High Cliff Point (Map 4).

High Cliff Point is a small, blunt, triangular headland pointing eastwards. Though it is the highest point on the section of coast of Grenada shown on the map its cliffs do not rise above 150 feet, and at the southern extremity are less than 25 feet high. South of this point for about ¾ mile the coast curves gently and regularly to form a bay. Inland, for as much as 600 yards at one point, the land is very low-lying, and

18

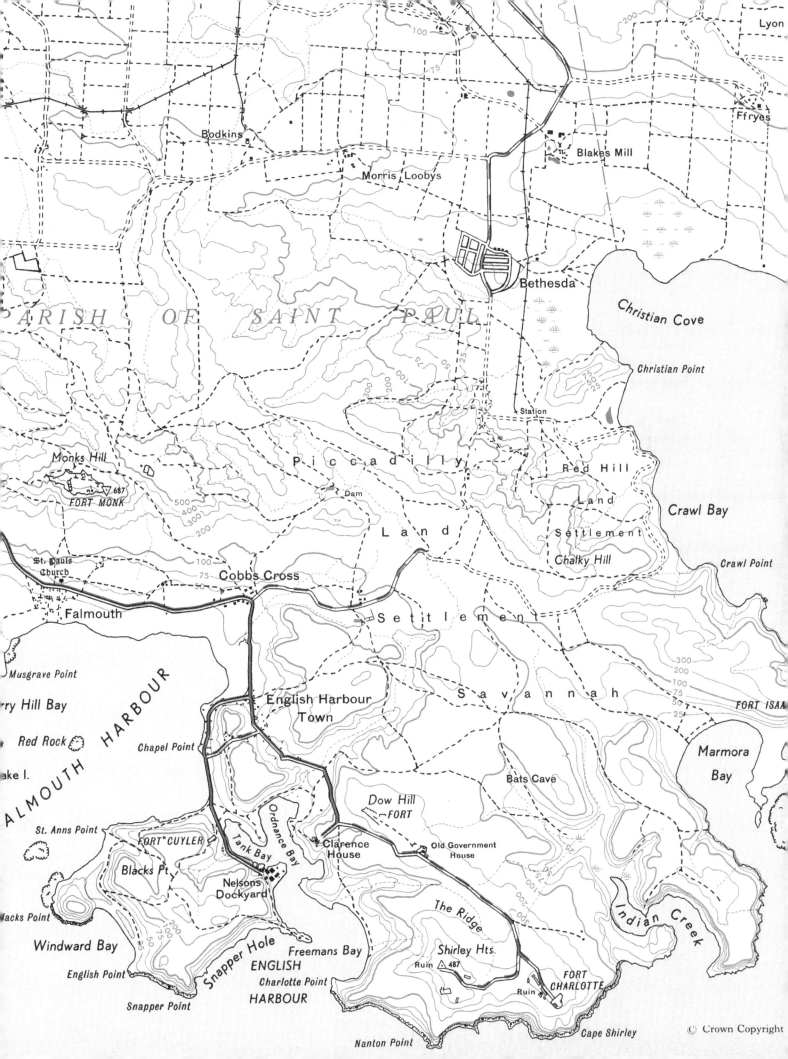

Lyon

Ffryes

Bodkins

Blakes Mill

Morris Loobys

PARISH OF SAINT PAUL

Bethesda

Christian Cove

Christian Point

Station

Monks Hill

Piccadilly

Red Hill

687

FORT MONK

Dam

Land

Crawl Bay

St. Pauls Church

Land

Settlement

Cobbs Cross

Chalky Hill

Crawl Point

Falmouth

Settlement

Musgrave Point

FALMOUTH HARBOUR

English Harbour Town

Savannah

FORT ISAA

Red Rock

Chapel Point

ake I.

Marmora Bay

St. Anns Point

Bats Cave

Dow Hill
FORT

FORT CUYLER

Tank Bay

Ordnance Bay

Clarence House

Old Government House

Blacks Pt.

Nelsons Dockyard

lacks Point

Indian Creek

The Ridge

Windward Bay

Snapper Hole

Freemans Bay

ENGLISH

Shirley Hts.

FORT CHARLOTTE

English Point

Charlotte Point

Ruin 487

Snapper Point

HARBOUR

Ruin

Nanton Point

Cape Shirley

© Crown Copyright

there are mangrove swamps in places. A small stream enters the bay near its northern end. Southwards again there is a well-marked change of about 20° in the direction of the coastline, and the material of which the beach is made is different.

Continue this description for another mile of coastline. Assignment 21

....

....

....

....

....

....

....

It is best—at least at first—to leave any analysis of physical features, such as coasts, until after you have described them fully. Then you can look back at your description and see what features need to be explained. ANALYSING COASTS

Looking back at the description of the coast shown on Map 4, certain questions suggest themselves. See if you can answer them. Assignment 22

(a) Why is there a headland at High Cliff Point? ....

....

(b) Why is it cliffed? ....

....

....

(c) Why is there a bay to the south? ....

....

....

(d) Why is the curve of this bay so regular? (Think where the sea may originally have extended to, and what could have happened since.)

....

....

....

....

(e) What is the beach material of the bay probably made of? ....

....

(f) Why do you think so? ....

....

....

(g) Give a possible explanation of the change in composition of beach material further south ....

....

....

....

Assignment 23
(Revision)

Imagine an aircraft coming in from the sea to land at Pearls Airport (Map 4)

(a) What bearing would the pilot follow? .... (b) How long is the airstrip he is about to land on? .... (c) If the aircraft is 500 feet above the sea when it is ½ mile from the end of the airstrip, can the pilot possibly see:

(i) Dunfermline Bridge? .... (ii) The Waterworks near Mount Horne Estate? .... (iii) The surface of Lake Antoine? ....

Assignment 24
(Revision)

Underline the correct answer in each of the brackets
(a) A gradient of 1 in 3 is about the same as a slope of (22°; 20°; 18°; 16°; 14°)
(b) A gradient of 1 in 5 is about the same as a slope of (17°; 15°; 13°; 11°; 9°)
(c) A gradient of 1 in 10 is about the same as a slope of (6°; 5°; 4°; 3°; 2°)
(d) A slope of 26° is about the same as a gradient of (1 in 1; 1 in 2; 1 in 4; 1 in 7)

DESCRIBING RIVERS AND THEIR VALLEYS

When describing a river—or part of a river—shown on a map, some of the things you should consider are its tributaries, its width, its gradient, and the size of its meanders. The shape and width of its valley are also important because they indicate the nature of erosion the land has undergone—and is still undergoing.

Here, as an example, is a description of the first 1,000 yards of the most northerly tributary of the Simon River (Map 4) beginning from its source a short distance south of Upper St. Johns.

The stream rises on the eastern slopes of a highland at a height of between 850 feet and 900 feet, and flows in an easterly direction in a small, steep-sided valley. For the first 600 yards (1800 feet) of its course, down to a height of 500 feet, its descent is steep and regular, the gradient being 1 in 5. Below this point its gradient is more gradual, being 1 in 12 for the next 400 yards.

The section drawn in Figure 12 across the upper part of this valley, over the ridge, and across the neighbouring valley to the south shows the nature of the landscape in this area.

Figure 12.

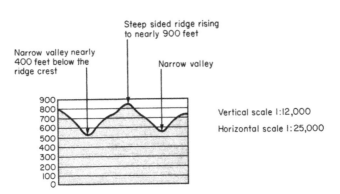

Steep sided ridge rising
to nearly 900 feet

Narrow valley nearly
400 feet below the
ridge crest

Narrow valley

Vertical scale 1:12,000

Horizontal scale 1:25,000

MAP 3 Part of Jamaica

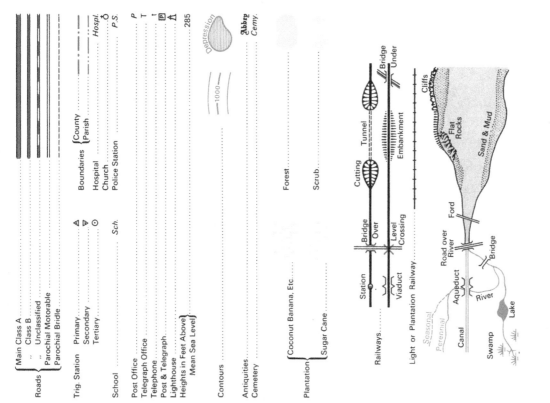

Roads { Main Class A / .. Class B / Unclassified / Parochial Motorable / Parochial Bridle

Trig. Station Primary / Secondary / Tertiary

School .... *Sch.*

Post Office .... P
Telegraph Office .... T
Telephone .... t
Post & Telegraph .... P
Lighthouse
Heights in Feet Above
Mean Sea Level } 285

Contours 1000

Antiquities ....
Cemetery .... *Cemy.*

Plantation { Coconut Banana, Etc. ....
Sugar Cane ....

Boundaries { County .... / Parish
Hospital .... *Hospl.*
Church ....
Police Station .... *P.S.*

Depression
Abbey *Abby*

Forest ....

Scrub ....

Railways .... Station / Bridge Over / Cutting / Tunnel / Bridge Under
Level Crossing / Viaduct / Embankment

Light or Plantation Railway ....

Seasonal / Perennial Cliffs
Flat Rocks
Sand & Mud
Road over River Ford
Aqueduct Bridge
Canal River
Swamp Lake

Scale 1:50,000

Linear Scale A

Continue this description of the river and its valley to Dunfermline Bridge. Use this space for a section of your own choosing. Remember to add a caption and a key.

Assignment 25

....
....
....
....
....
....
....

22

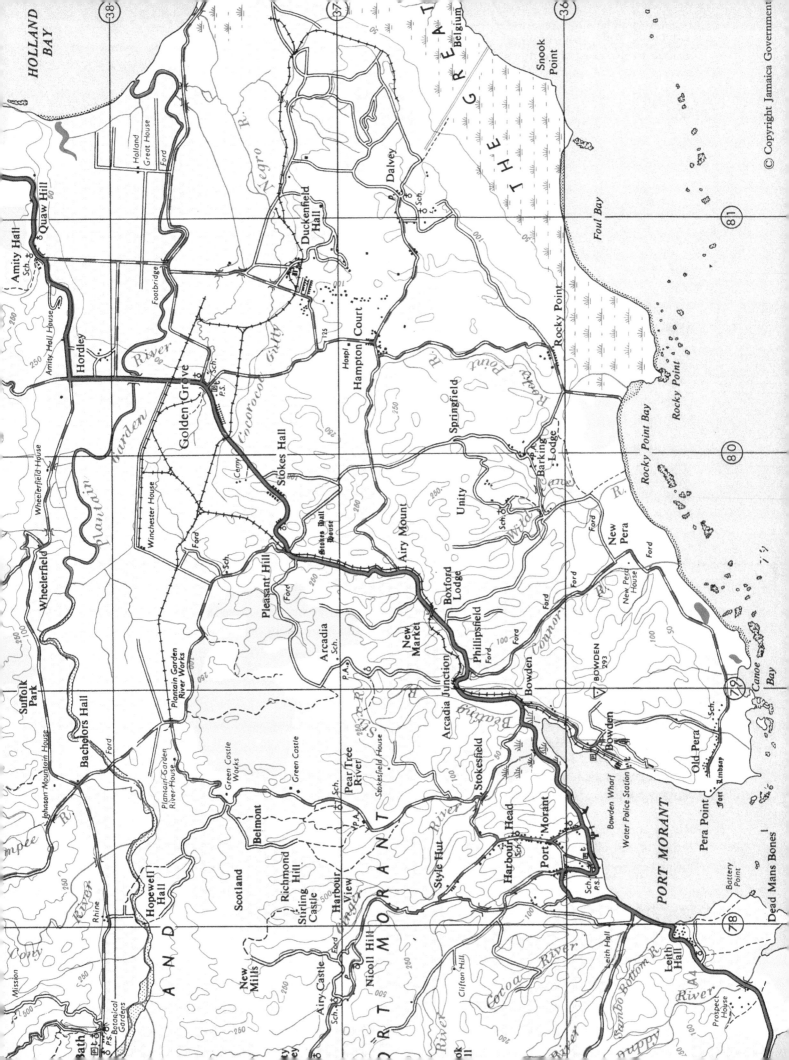

Conclude your description of the Simon River and its valley from Dunfermline Bridge to the coast. Use this space for another section.

Did you remember to add a caption and a key?

....

....

....

....

....

....

....

....

....

The crest of the land separating one river and its tributaries from another is called a *watershed*. In mountainous areas watersheds are rugged, winding, and difficult to follow. They form natural boundary lines between parishes, counties, states and nations. In areas of gentle relief they may form natural routeways and be followed by roads.

On the maps themselves, mark the approximate lines of the following watersheds: Assignment 26
(a) The watershed between the Farm River (Map 1) and the streams further north.
(b) The watershed from Monks Hill to Fort Isaac (Map 2).
(c) The watershed between the Plantain Garden River (Map 3) and the streams flowing southwards.

Rivers and their valleys change so much from their sources to their mouths that in order to simplify description geographers divide them into upper, middle and lower courses.

See what your physical geography textbook says about the upper middle and lower Assignment 27 course of a 'typical' river. Choose three characteristics of each, write them in Table 3 on page 25, and say whether they are present or absent in the Simon River.

	Three characteristics of a typical upper course of a river	Present or absent in the Simon River
1		
2		
3		
	Three characteristics of a typical middle course of a river	Present or absent in the Simon River
1		
2		
3		
	Three characteristics of a typical lower course of a river	Present or absent in the Simon River
1		
2		
3		

(a) (i) Draw a section from the main road west of Lake Antoine (Map 4) due east across the middle of the lake to the coast. What must you remember to add above and below your section?

Assignment 28 (Revision)

```
500
400
300
200
100
  0
```

(ii) What is the probable origin of Lake Antoine? (Map 4) ....

(iii) Why do you think so? ....

....

....

(b) (i) Describe one mile of the course of the Plantain Garden River (Map 3) south

of Bath...

....(continue your description on page 26)

MAP 4 Part of Grenada

Main Roads ....
Culvert

Secondary Roads ....

Other Roads ....
Metalled

Tracks and Footpaths ....

Named Buildings .... ■

Other Buildings ....

Police Station .... ■ Pol Sta

Church, Chapel, School .... ■ Ch ■ Cha ■ Sch

Hospital, Health Centre .... ■ Hosp ■ HC

Water Tank .... ■ WT

Trigonometrical Stations:- *Main* .... △

„   „   :- *Minor* .... ▽

Heights in feet given to ground level .... ·1823

Contours V.1.50' (Below 250' – 25') .... —300— —250— —225—

Windbreaks ....

Mangrove .... ^ ^ ^

Swamp ....

Cultivation and Plantation (S = Sugar, C = Coconut)

Ruins, Antiquity ....

Watercourses
Sand or Mud
Quarry
Pond
Dam
Waterfall
Rapids
Sand Dunes
Bridge
Lake
Ford
Flat Rock
Rapids
Ferry
Sand or Mud
Lighthouse
Waterfall
Cliffs
Steep Slopes
Boulder Rock
Outcrop Rock

Scale 1:25,000

Linear Scale A

```
1 Mile    ¾      ½      ¼      0                              1                              2  Mile

      Yards  1000           0           1000           2000         Yards  3000
```

Linear Scale B

```
1 Kilometre          0              1              2              3  Kilometres
```

....
....
....
....
....
....

(ii) What indication is there that its volume varies considerably?

....

(iii) Describe the last mile of its course before it enters Holland Bay.

....
....
....
....
....

26

Describing a total landscape

Now that you have had practice in describing coasts and valleys the next step is to describe the total landscape shown on a map. One of the best ways of doing so is to sub-divide it into 'physical regions', that is into areas throughout which the relief is similar. Each of these regions can then be described and, afterwards, compared and contrasted with one another. For example on Map 3 the valley of the Plantain Garden River can be chosen as one region, the swamp as another, and the uplands between them as a third.

Describe the third region (i.e. the uplands shown on Map 3) including in your description their length and width, their elevation, their slopes, their distance from the sea, and their drainage.

Assignment 29

....

....

....

....

....

....

....

....

....

Remember there is no hard and fast rule about choosing regions. Your choice will depend partly on the degree of detail in your description and partly on the purpose you have in mind. For example, the coastlands between Pera Point and the swamp (Map 3) may be considered as part of the coastal plain in that part of Jamaica or as a small region in its own right.

(a) (i) Describe the swamp shown in Map 3 ....

Assignment 30

....

....

....

....

(ii) Describe the coastal strip between the swamp and Pera Point.

....

....

....

....

(iii) In what ways are they similar? ....
....
....
(iv) In what ways do they differ? ....
....
....

(b) Divide the land shown on Map 2 into physical regions (choose about six) and name them carefully so that they are readily distinguishable from one another.

(i) ....

(ii) ....

(iii) ....

(iv) ....

(v) ....

(vi) ....

Another way of describing landscape is to choose a small area which is typical of one physical region and describe it. You can then compare it with other small areas chosen as being typical of other regions.

(a) Describe the land shown in square 7M (Map 1) under the following headings: Assignment 31

(i) The height of the land ....
....
....
....

(ii) The slopes (and gradients) ....
....
....

(iii) The streams ....
....
....
....

(iv) Their valleys ....
....
....
....

MAP 5 Part of Barbados

Culvert

Roads – *Main* ....

„ – *Secondary* ....

„ – *Other* ....

Tracks & Intervals ....

Footpaths ....

Boundaries – *City* ....

„ – *Parish* ....

Antiquities .... **𝔉ort**

Chimney .... *Chy*

Church   Chapel   School .... Ch   Cha   Sch

Sinkhole, Well .... O SH   O W

Gully or Watercourse ....

Pond, Pond seasonal ....

Sand or Mud ....

Quarry, Cliff ....

Trigonometrical Station – *Major* .... △

„   „   – *Minor* .... ▽

„   „   – *Subsidiary* .... ⊙

Heights in feet above Sea Level 546

Contours *(Vertical Interval 20 ft.)* ....

Depression ....

Steep Slopes ....

Trees – Deciduous ....

„ – Casuarina ....

„ – Palm ....

Scrub and Sour Grass ....

Sugar Cane ....

Grassland ....

Scale 1:10,000

Linear Scale A

Yards   1000   500   0   ¼   ½   Mile

Linear Scale B

500   Metres   0   ¼   ½   ¾   1   Kilometre

(b) Describe the land shown in Map 1 square 10I in much the same way.

....

....

....

....

....

....

....

....

(c) Why is the land shown in square 7M so high and so steep?
NOTE: The word 'Soufriere' is a clue. For its meaning see *Caribbean Lands*.

....

....

....

Some geographical relationships

Relief is a major factor influencing the location of roads. For example, it is always expensive and sometimes impossible to build roads in mountainous areas, so they tend to avoid peaks and ridges and follow instead any passes that exist between them. In contrast, lowlands present few obstacles to road building unless they are swampy.

River valleys commonly provide natural routeways and are therefore often followed by roads. Rivers themselves, however, are obstacles to road transport and they are crossed only at key points. Coasts are an absolute barrier to roads.

(a) Describe the relationship between the relief and the main road shown in square 9J on Map 1 ....

....

....

....

....

....

....

(b) Describe and explain the reason for the hairpin bend on the main road west of Lake Antoine (Map 4) ....

....

....

....

(c) Examine the course of the Plantain Garden River (Map 3). In how many places can it be crossed by wheeled traffic? ....

How many of these are sure to be closed when the river is in spate? ....

How far is it from Wheelerfield 796383 to Winchester House 796379 (Map 3)

(i) in a straight line? .... yards, or .... metres.

(ii) by road? .... yards, or .... metres.

(d) The road from Bath through Wheelerfield to Hordley runs some distance away from the river. Why do you think it is not nearer? ....

....

(e) Describe and account for the pattern of roads east of Pera Point and south of grid line 36 (Map 3) ....

....

....

....

....

....

....

....

....

....

....

....

....

A close relationship often exists between relief and the way the land is put to use. Common examples in the West Indies are forests on high mountainsides, sugar cane on flat land, and coconuts close to the sea.

RELIEF AND LAND USE

(a) Draw a section along part of the parish boundary on Map 5 (page 31) to show the relief of the land above 80 feet. Use a vertical scale of $\frac{1}{10}$ inch = 40 feet.

Assignment 33

(b) Indicate in detail on your section the ways in which the land is being used.

(c) What relationships between relief and land use are revealed by your section?

....

....

....

....

....

....

....

....

(d) Look at the section you have drawn. Does it include a caption and a key?

(a) What is the upper limit of cultivation in that part of Grenada shown in Map 4?

Assignment 34

....

MAP 6 Part of Guyana

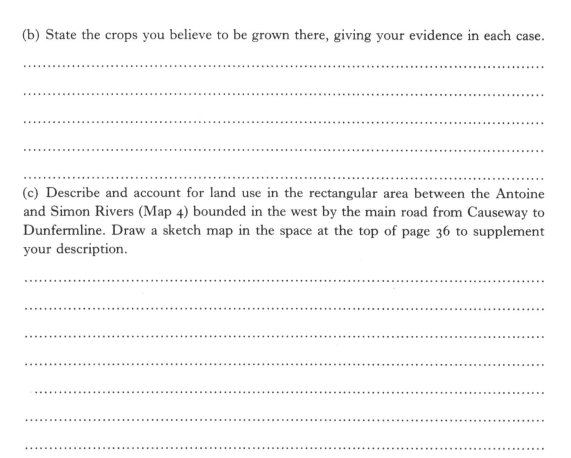

Telegraph or Telephone
Lines ....
,, along Road ....
,, along Track ....
Post, Telegraph Office ....■ P ■ T
Police Station ....■ PS
Hospital ....+Hosp
Church, Mission ....+Ch+M
Mosque, Temple ....
School ....■ Sch
Barracks ....■ B
Spot Heights in Feet ....·5653
Light Forest ....
Scrub ....
Marsh or Swamp ....
Courida Bush ....
Rice ....
Palm, Mango ....
Sago, Bamboo ....

Towns or areas with
permanent buildings ....
Other populated
areas ....
Villages, Isolated
buildings ....
Main Roads ....
Secondary Roads ....
Tracks ....
Minor Tracks ....
Railways with Station,
Siding ....Sta
,, Light ....
Drain or Irrigation
Chan. and Track ....
Drain or Irrigation
Channel ....D
Power Transmission
Lines ....
Wireless Station ....

Scale 1:50,000

(b) State the crops you believe to be grown there, giving your evidence in each case.

....
....
....
....

(c) Describe and account for land use in the rectangular area between the Antoine and Simon Rivers (Map 4) bounded in the west by the main road from Causeway to Dunfermline. Draw a sketch map in the space at the top of page 36 to supplement your description.

....
....
....
....
....
....
....

Linear Scale A

Linear Scale B

GEORGETOWN

....

....

....

....

....

....

Draw your sketch map in the space below.

Roads are often a good clue to the density of settlement of an area and the degree **LAND USE**
to which the land is profitably used. For example, dense road networks always **AND ROADS**
exist in towns and their suburbs. There are also many second-class roads in rich
agricultural districts so that the produce grown there can be taken to its market.
Tracks—often very busy ones—may take the place of roads in areas of difficult
terrain.

Main roads are sometimes an exception. They join major population centres and
may run through areas of little importance. However, their very existence attracts
settlement which tends to be strung out in a long narrow line: a pattern known as
'ribbon settlement'.

(a) On the fragment of the Parish of St. Michael shown in Map 5 colour the main Assignment 35
road red, the secondary roads orange and the other roads yellow.
(b) What relationships can you find between

(i) roads and relief? ....

....

....

....

....

(ii) roads and land use? ....

....

....

....

....

Assignment 36 (Revision)

(a) What is the shortest distance by road from the Ocean View Hotel (Map 5) to the Club House on Rockley Golf Course? (i) To the nearest mile .... (ii) In yards .... (iii) To the nearest km. .... (iv) In metres ....

(b) Assuming the Club House to be 50 feet above sea level, about how high would the masts north of Highway 6 have to be, to be visible from it? ....

(c) Describe the coastline shown on Map 5 ....

....

....

....

....

....

....

....

....

....

....

....

MAP 7 Part of Trinidad

Scale 1:10,000

Railway – Double	
" – Single	
	Siding
Road – 1st Class	
" – 2nd Class	
" – 3rd Class Paved	
" – 3rd Class Unpaved	
" – 4th Class	
Footpath	
Power Transmission Line	
Public Building, Other	
Contours (V.1 25 feet)	250 225 200
Rice	
Line of Trees	

Forest	
Positioned Tree	
Grassland	
Sugar Cane	
Cemetery	Cem □
Warden's Office	WO ■
Church	Ch ■
School	Sch ■
Police Station	PS ■
Post Office	PO ■
Temple	Tem ■
Sports Ground	SG ■

Linear Scale A

Linear Scale B

(d) State three ways in which the coast shown on Map 5 differs from that around Port Morant (Map 3).

(i) ....

(ii) ....

(iii) ....

(e) Suggest reasons to explain the contrasts in the economic development of the two coastlines...

....

....

....

....

....

....

....

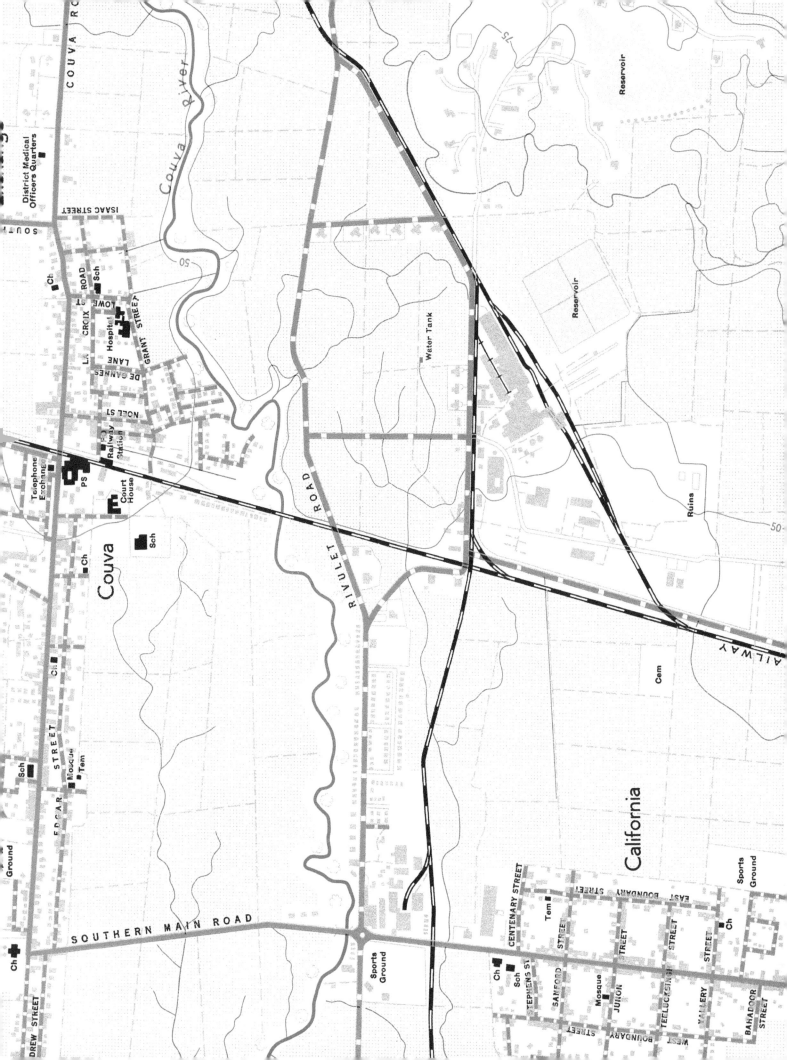

Settlements

Maps often give a clear picture of the site and shape of towns, villages, and other settlements, and provide evidence of some of the reasons for their existence.

(a) Colour all the built-up area (including roads) shown on Map 6 red. Now colour all the agricultural areas, making a distinction between sugar cane and coconuts.

Assignment 37

(b) What, approximately, is the area of Georgetown? ....

(c) Given that its population was 120,000 at the time the map was made, what was the population per square mile? ....

(d) Describe the site and shape of Georgetown ....

....

....

....

....

(e) What evidence is there that the Demerara River is important for shipping while the sea coast is not? ....

....

....

....

(f) How can you tell that the sea coast is shifting while the banks of the Demerara River are stable? ....

....

....

....

(g) How does this explain the importance of the Demerara River for shipping?

....

....

(h) In what ways is the Demerara River a barrier to communications and not only a routeway? ....

....

....

(i) Name three functions of Georgetown, giving supporting evidence.

(i) ....

....

(ii) ....

....

(iii) ....

....

(a) Describe what would you see if you went from the sugar factory shown near +85   Assignment 38

on Map 6:

(i) To the sea along a bearing of 17° ....

....

....

....

....

....

....

....

....

(ii) To the river along a bearing of 287° ....

....

....

....

....

....

....

....

(b) Describe the view you would see from an aeroplane over Houston looking towards

Vreed En Hoop (Map 6) ....

....

....

....

....

....

....

....

Check your answer against the photograph shown on page 181 of *Caribbean Lands*
(3rd edn).

By using large-scale maps it is sometimes possible to make a rough estimate of the number of people living in an area and deduce something of its history and present-day society.

(a) About how many houses are there in California (Map 7 on page 39)? .... Assignment 39
(Remember to allow for those that are joined together.)

  (i) Assuming that 5 people live in each house, about how many people live in

    California? ....

  (ii) What is the area of California? ....

  (iii) What do you calculate the population per square mile to be? ....

(b) (i) In what way do the street names suggest that California was originally a

    planned settlement? ....

    ....

    ....

    ....

  (ii) Given that the large building shown in the south centre of Map 7 is a sugar factory, for what purpose do you think California was built?

    ....

    ....

  (iii) For whom do you think it was built?..

  (iv) Why do you think so? ....

    ....

    ....

  (v) What evidence is there that the descendants of these people live there today?

    ....

    ....

  (vi) What indications are there that California has grown since it was originally

    built? ....

    ....

    ....

    ....

  (vii) Can you suggest another change that seems to have occurred?

    ....

(c) (i) How many houses are drawn along the 3rd class unpaved road near the

    reservoir shown in the south-east corner of Map 7? ....

  (ii) Would you say that the population density here was (a) about the same (b) less

    dense (c) much less dense than California? ....

(iii) Name two differences between these houses and those in California.

....

....

(iv) Name two other differences between this settlement and California.

....

....

....

(v) What do you suggest is the function of this settlement?

....

(c) Describe and account for the settlement pattern leading into Couva from the

west ....

....

....

....

....

....

....

(d) Describe the site and shape of Couva and say what functions you think it serves.

....

....

....

....

....

....

....

....

(e) Describe the view from the south-east corner of Map 7 over the reservoir and

the sugar factory ....

....

....

....

....

....

Compare your answer with the photograph on page 67 of *Caribbean Lands* (3rd edn).

Drawing sketch maps on a reduced scale

The purpose of a sketch map is to display, in a simple and vivid way, a few selected features of an area so as to illustrate a few specific points. It should not attempt to serve as a scaled-down version of an original map. Usually, therefore, in drawing a sketch map only an approximate outline is necessary. But this is not always sufficient. If, for example, a series of relationships is to be depicted on a series of maps (say roads and relief on one map, roads and agriculture on another, and roads and settlement on a third) the road pattern on the maps should be reasonably accurate. (Only the first needs to be carefully drawn because the others can be traced from it.)

Here is a way of drawing a reasonably accurate outline on a reduced scale.

With a pencil, draw a grid of squares over the original map. — Step 1

Draw a similar pattern of smaller squares on a piece of plain paper. If, for example, the scale of your map is to be half that of the original, the outline of your map and each of the squares you draw must be half the length and half the height of the original. — Step 2

Using the squares as a guide, outline the particular features you wish to represent. — Step 3

Add a title, a key, and a simple linear scale showing distances in miles or in kilometres. — Step 4

(a) Turn to Map 1 (page 15) and look at the grid of 30 squares between 7 to 11H and 7 to 11M. Draw a similar grid, at half the scale, inside the rectangle on page 45. — Assignment 40
(b) Mark with faint pencil dots the points where roads cross your grid lines.
(c) Draw in the main motorable roads and colour them *red*.
(d) Repeat processes (b) and (c) so as to draw in the 1,000 ft. and 500 ft. contours. Colour the land above 1,000 feet *brown*, the land between 500 and 1,000 feet *yellow*, and the land under 500 feet *green*.
(e) What relationships are shown on your sketch map between relief and roads?

....

....

....

Compare this with what you wrote in Assignment 32(a) (page 32).
(f) Add a title and a key to your sketch map, and complete the two linear scales.

(a) On a sheet of paper draw a rectangle 5·6 in. by 4·4 in. and draw a reduced version of Map 3 to show (i) the river systems, (ii) the main watershed, (iii) the roads, and (iv) the regions you listed on page 29. — Assignment 41
(b) Add a title, a scale, and a key to your map.

Geography and planning

A knowledge of geography is essential for those who are planning new housing projects, communications, water supplies, and other developments. In the following assignment you can use your skill in map interpretation in a variety of imaginary planning situations.

Linear Scale A └──┘ miles

Linear Scale B └──┘ kilometres

(a) Imagine you are planning a new middle-class housing settlement for about 5,000 people somewhere in that section of Jamaica which is shown on Map 3 (page 23).

Assignment 42

(i) Outline in pencil the area you would select for its location.

(ii) Give reasons for your choice ....

....

....

....

....

(iii) In the area you have outlined, show by means of conventional symbols the various community services which you believe would be needed there.

(iv) Choose one member of your class to act as Town Planner and seven other members to serve as a Planning Committee to consider his proposals for locating the new housing settlement. In the next class period allow the Town Planner five minutes to address the Committee and ten more minutes for the Committee to question him and to hold a secret ballot on whether to accept or reject his proposals. Now, by a show of hands, vote on whether you think the Committee should have accepted the proposals or not. Compare your class vote with that of the Committee. Discuss the results.

45

(b) Imagine that a new hotel complex is to be built between Marmora Bay and Indian Creek (Map 2, page 19). Plan a new main road connecting it to the north-west corner of the map, which you may assume is on the way to the capital (St John's).
(i) Draw on the map itself the route you would choose for this road.
(ii) Give reasons to say why you chose this particular route ....

....

....

....

....

....

(iii) Discuss the routes chosen by several members of the class and see if you can come to any agreement about the best route.
(c) Imagine that you had to choose the site of a dam to create a reservoir which could supply gravity-fed water to the majority of the people living in that section of Grenada which is shown on Map 4 (page 27). You are told that in order to conserve funds you should take into account the need to supply as many people as possible with the shortest possible pipeline.
(i) Mark the place where you would build the dam.
(ii) Outline the area that would be covered with water when the reservoir is full.

How high is your dam? ....
(iii) On the map itself shade all the places which could not be supplied with water directly from your reservoir.
(iv) In this space draw a section along the line you would propose for one of your pipelines.

(d) Imagine that you had to choose the site for an airfield with an airstrip 1,200 yards long somewhere in that section of Jamaica which is shown on Map 3 (page 23). Take into account such things as elevation, prevailing winds, access by road, approach routes for planes and the undesirability of having to find new accommodation for large numbers of people affected by the proximity of the new airfield.
(i) Draw the position of the airfield on the map.
(ii) Give reasons for your choice of location ....

....

....

....

(iii) Discuss the sites chosen by several members of your class and see if you can come to any agreement about the best site.

Answer these questions, using Map 5, page 31.

(a) What is the compass bearing of the Central Livestock Station from Ventnor?......

(b) What is (i) The straight line distance between the two points? .... yards

.... metres.

(ii) The shortest distance by road? .... yards .... kilometres.

(iii) The difference in height? .... (iv) The gradient? ....

(c) From Ventnor, what are the compass bearings to (i) Las Palmas? ....

(ii) The Royal Hotel? .... (iii) The Ocean View Hotel? ....
(d) Divide the map into quarters by ruling one line from top to bottom and another from side to side.
(i) On a piece of plain paper draw a sketch map to show roads and settlement in the north-east quarter of the map.
(ii) How do the roads and settlement pattern in this area compare with those shown

in the south-west quarter? ....

....

....

....

(e) On the map itself, mark (i) Three good locations for a gas station; (ii) A good location for a shopping plaza (taking into account the space needed for parking); (iii) A good location for a large sports stadium; (iv) A good location for an infant school.
Discuss your choices in class.

Answer these questions, using Map 1 (page 15) and Map 3 (page 23).

(a) Calculate the approximate area of the land shown on Map 1

(i) In square miles .... (ii) In square kilometres ....

(b) Explain how you did the second calculation...

....

....

(c) How many miles of motorable road are there for each square mile of land? (You can calculate the answer by using the figure you gave in Assignment 11(c).)

....

(d) Calculate the approximate area of the land shown on Map 3 .... sq. miles

.... sq. km.
(e) As the two *maps* are the same size, explain why there should be such a difference

in the land areas shown on Maps 1 and 3 ....

....

(f) About how many miles of motorable road are there for each square mile of land

shown on Map 3? .... Explain the contrast with Montserrat. ....

....

47

Answer these questions, using Map 6, page 35.

(a) What bearing does the ferry follow from the pier at Vreed en Hoop to George-town? .... and back again? .... How far is the double journey? ....

(b) How long are (i) the river banks? .... miles .... km. (ii) the coastline? .... miles .... km. (iii) the railway? .... miles .... km.

(c) Say where you would choose to locate *either* a cement factory *or* an oil refinery and give reasons for your choice of site ....
....
....
....

Assignment 46
(Revision)

Answer these questions, using Map 2, page 19.

(a) Calculate the approximate area of land shown on the map .... sq. miles.

(b) On Map 2 itself mark the highest point. Describe the view looking south. ....
....
....
....

(c) Imagine you are standing on the 100 foot contour just beyond the end of the foot-path which leads around Freemans Bay towards Charlotte Point. (i) Describe the view you would see in a north-westerly direction ....
....
....
....

(ii) Is Monks Hill visible from this point? .... (iii) Say how you know.
....

(d) Using the map and the photograph, what indications can you find that, in spite of its location beside the sea, Nelson's Dockyard is not an important port?
....
....
....
....
....

On a separate piece of paper draw a reduced version of Map 7, page 35, outlining and naming about six sub-regions.

Assignment 47
(Revision)